1

YOUR WEBSITE

Gets referrals.

Graeme Smith

PUBLISHED ON AMAZON.com
by
LABYRINTH BOOKS

DEDICATION:

This book is dedicated to my family.
      Hele-ly (Ly).
            my wife:

    Ingrid.
            our daughter:

    Marie.
            my former wife:

    Fiona, Natalie and Michael
            our children:

    Georgie
            Michael's wife:

    Pearl, Kiki and Martha.
            their children:

They have had to put up with me for many years and I thank them for that.
I hope this book gives them an insight into what has occupied me much of the time.
They have all achieved worthwhile and interesting careers in the absence of much help from me.
I congratulate them for their achievements.

THANKS:

I greatly appreciate the contribution made to this book by comments and suggestions from:

Mike Barr – Adelaide, Australia

Richard Bruland - Los Angeles, USA

Tracey Creighton - Merimbula, Australia

Evelyn Dunphy – Maine, USA

Geoff Fellows – Wagga Wagga, Australia

Michelle Grace - Brisbane, Australia

Leanne Halls – North Sydney, Australia

Heidi Jeffries – Ferny Hills, Australia

Kathy Kay Voysey - Mudgee, Australia.

Vince Miller – 'Australian Artist' and 'International Artist'.

John Newell - Ontario, Canada

David Voigt – Yarramalong, Australia.

HOW TO USE THIS BOOK.

First think - then do.
Usually people don't think through things to the level they need to.
Because of that, they have projects instead of tasks on their "to do" list.
That leads to procrastination as it hasn't been broken down to a task level.

So go through your book once to understand it.
Go through it again.

Then start at the idea you would like to implement first.
Make notes of the steps you will need to take and the resources required.
Use these notes to create a step by step system for implementing the guide.
Often you will not refer back to an original, as you've created YOUR system.

The first question you ask and answer is "Why is this being done?"
How does this align with where you want to get to?
What are the strategic implications of doing this?
Does this fit with getting to a goal in the shortest and fastest amount of time?

What would it be like if it were totally successful?
Define it - what is success for this project and how will you know?

Now brainstorm all the tasks that are involved in your project.
It's important not to go linear too fast with this.
By linear, I mean step one, step two, step three, and step four.
You end up cutting off options.
Plan step one, two, and three, a specific step that might be number four.
If you start steps quickly, other ways of one, two or three may not appear.

The first third of any brainstorming session is really easy.
Just come up with lots of ideas.
The second third is challenging - go through ideas and see where they lead.
Then push yourself to think a little bit outside the box.
That's often where the big idea is!

That's where the most powerful way to get a project done fastest - is.
Most people never get to that level and end up short-changing themselves.
Then their project takes longer and they also procrastinate.
This final brainstorming part of the equation is incredibly important.

Once you fully brainstorm put your options into a linear sequence.
Then you can figure out what you've overlooked.
Everything becomes obvious as you get your tasks in order.
Now add missing steps and you have laid out your task list for this project.

Once you've organized the tasks into a linear process decide:
What things can you start immediately?
What can be started that is not dependent on things to occur before them?
Obviously that is step one.
Step five or six or twenty that don't really rely on anything else to get done.
You can get started on them right away too.

Now use a folder.
Write things you think of at the time and also cross off things as you do them.
Add in stuff that is relevant from time to time.

WHAT IS MARKETING?

Marketing is the process of finding buyers AND making sales
It is exactly the same process no matter what is being sold!
In some cases the process is simple like selling apples at a roadside stall.
It can be very complex like selling aero-planes for a government's air-force.
Most, including selling artworks, is somewhere in between these poles.

Think about fishing and you'll understand marketing.
Does a fisherman catch anything out in a desert?
NO, for there are simply no fish there.

You must market where there are possible buyers.
A fisherman must go where the fish are – where there is water.
That's a start but there are still no fish in a swimming pool are there?
They need to be in the right kind of water – a river, lake or at sea.

But different fish swim in different waters!
Sharks and marlin are in the ocean, while bream live mainly in rivers.
Likewise you must know who you are targeting with your marketing.
Will it be businesses, first home buyers, investors or what?
Each will need a different marketing program.

OK you are now in the right water for the kind of fish you are after!
Some species are nocturnal and they will not be caught during the day.
Your marketing needs to be when the target is likely to be most receptive.
Will it be at work, nights or weekends?

You are at the right place and time so how do you catch the fish?
Usually you'll have a fishing rod.
Is it the right kind for the fish you want to catch?
You won't catch a shark with the kind of rod that takes a trout!
Your marketing must be attractive to the people you are after.

Do you have the right bait?
Again different bait attracts different fish.
A carcass for the shark but just a worm for many other species.
Can you provide something that your target market will find attractive?

But throwing any bait into the water catches nothing at all!
The bait must be attached to a hook.
Without the fish taking the hook there is no catch.
Different hooks are needed for different kinds of fish.

Different hooks are also needed for different markets.
The right hook gets your market to take the next step to a purchase.
But this only needs to be a little step.
But hooks only catch the fish, they're still in the water.

Not your boat or beach, unless the hook is on the end of a line.
What is your line like, is it strong enough you the fish you are after?
Again this varies for the kind of fish.

How do you get your prospect to seriously consider what you sell?
For someone buying a print it will not need to be sophisticated.
But selling an original Renoir will be considerably more complicated.

That still no fish for the line has to have a reel for that to happen.
Again different reels for different fish.
The right reel allows you to bring the fish to the end of your fishing line.
But it's still not in the boat is it?

You must lift the fish out of the water into your boat or onto the beach.
Fishing nets do this so now you have your catch.
The fish is yours to do what you want with.

You can even sell the fish but who might want to buy?
It could be someone who sells fish for food or live for a fish-tank or pool.
They could even be for re-stocking natural water places.

Where can you find them?

You must look where the fish buyers are!

Follow the path of the fisherman.

And eventually you have a prospect asking can they buy.

You have made a sale AND you can make more sales the same way.

Making the sale is a five step process.

In order you work from the top through to the bottom group.

> **SUSPECTS** are people who possibly want what you have for sale.
>
> **PROSPECTS** are people likely to want what you have for sale.
>
> **BUYERS** are those who have bought what you are selling.
>
> **REPEAT BUYERS** continue to buy what you sell.
>
> **ADVOCATES** help you sell to others.

The reverse sequence is the order of importance to your sales.

INDEX: REFERRAL WEBSITE

Acknowledgement:

Evelyn Dunphy provided the spark that triggered this chapter.
I am very grateful to Evelyn for her ideas.
Evelyn is from Maine, USA.

Chapter One: Develop an effective website.

1. The start stops most people.
2. Branding is what big companies do.
3. Professional artists sell artworks to people who want to buy.
4. Asking questions can handle any sales situation.
5. How can you create better-performing marketing?
6. Testimonials can be very powerful and persuasive!
7. A client is anybody who wants to spend money on your work.
8. The best clients are the clients you already have.
9. Are you status conscious?
10. Many products gain status through pricing.
11. Is word of mouth advertising best?

1. The start stops most people.

Back in my art student days I learnt how to deal with this problem.
I was in the third year of a four-year art course.
One memorable evening I arrived for his class to find I was the only person!
I wasn't happy but decided not to waste my time.
I looked at my blank canvas and didn't know what to do.
Yes the start certainly had me stopped.
Have you had this feeling?

If I was going to waste paint it may as well be a colour I didn't like!
So I mixed Indian red with turps, still with no real idea about what to do.
I still didn't know what to do though.
Then just because the paint was runny, I flicked some onto the canvas.
I splashed some more and still further splashes followed the first ones.
It was a bit like an Indian red 'Jackson Pollock'.

Still not knowing what to do, I decided to join up the dots.
After a while, I looked again at the canvas, all covered with irregular shapes.

But I was no further ahead in working out what to do.
More speculation led me to an idea that I could colour some of the shapes.
I used white paint and coloured some shapes with it and Indian red mixtures.
Eventually I finished and had another look at what I had done.

Even though I didn't know what to do, I had done something.
It was an abstract painting.
I hadn't anticipated it before starting but was pleased with the result.
The next week the same thing happened.
I shut my eyes made random marks and again made an abstract painting.
I continued like this for the rest of the year for no longer did the start stop me.
I had found out the solution to that particular artistic problem.

It doesn't matter what you do at the start, as long as you do something.
It can be quite random although that doesn't necessarily have to be the case.

Starting a new website is no different!
Don't wait, just get on with it.
Write down whatever you can think of.
Add any other details as well.
Just get started!
Now your imagination can work for you!
You'll think of more details and over time you will develop your site.
It will be better than you could ever have imagined, but you do have to start.

2. Branding is what big companies do.

But this doesn't mean small business and even artists, cannot do it too!
Any business can gain considerable advantages from successful branding.

But just what is branding.
It happens to cattle on a ranch or farm.
Cattle receive a very painful mark on their rump which stays for life.
It's a symbol showing the cattle belong to a particular property.

A brand shows something belongs to an organization or individual.
I'm sure you are familiar with symbols that represent peace or women.
Even small companies have logos that that is what the brand symbols are.

How about artists?
Well our signature on each work is our logo and our name is our brand.
You know who painted a painting as soon as you read 'Vincent'.

There's no need to even study the work itself!
Your brand is based on your works, pricing, promotion, and appearance.
All you do to establish and foster an artistic career combines to form a brand.
It's your unique way of doing things.
Find the essential core and promote it by linking it with your logo (signature).

Your brand is how people think about you.
Your logo is on all works, including those not done yet.
A brand symbol is a powerful marketing advantage in this type of promotion.
You need to sell yourself to sell your business.
You are the key person in your business.
Selling yourself is not easy for many people, but it can be learned.
People in other careers do it so there's no reason to think artists are different.

The best way to communicate a brand is via word of mouth.
Because most consumers, including art buyers are cynical about advertising.
BUT they trust what friends and colleagues tell them, often they shouldn't.

That's how it works in the art-world too.
People buy on the recommendation of friends, reputation, and experience.

Building a brand is not rocket science.
It's a matter of making sure individual marketing strategies are integrated.

Then communicate your brand message to your target market.
The brand message is what your works or services can do for them.
Marketing people say that everything is branding and branding is everything.
They're right if you assume that a business is its marketing.
Be jealous guarding against anything reducing the effect of a brand.
Such as placing inferior works on sale.

Don't overlook the obvious.
You send, or even pay, those monthly invoices.
Suggest buying ideas based on purchase history of the person contacted.
Use this as a way of sharing good news about your career and what you do.
But keep your ego under control which for some artists is very difficult.
After isolation of the studio they just love the public aspects of their career.
It's easy to go broke with functions for existing, potential clients and friends.

Resist the temptation to invest in advertising.
If all you want is to see your name and works prominently displayed.
Often this is called getting exposure and is a waste of time as well as money.
An exhibition is to sell works not just show them and a website is similar!
But branding is not advertising.
Advertising can be used to establish and promote the brand.
BUT every business has a brand, even without advertising.

3. Professional artists sell artworks to people who want to buy.

Most artists assume that the function of marketing is to bring people in.
Bring them to the gallery, studio or wherever you sell your works or teaching.
Then salesmanship takes over and possibly this is close to the mark too.
But let's take that idea a bit further.

Marketing is to attract clients who keep coming back for what you sell.
Many business people, and probably all artists, don't understand this.
They try to make new sales every day, instead of focusing on longer term.
Once you bring a person into your sphere of influence, never let them go!

The best way to do this is to target this kind of prospect from the start.
Business people forget the investment needed to get someone in their door.
Ten people coming in has cost something, whether they buy or not.
The fact that these people are there, shows interest.
They need and probably want to develop the relationship.
That's the beginning of becoming a client or long-term client.

The more people keep coming back the more profitable your career is.
This is not because you charge more or anything like that.
One of the main costs in a business is attracting a prospect in the first place.
The cost of re-selling, over and over again, is small.

It costs very little to keep someone coming back.
A smile, outstanding service, a thank you, a friendly face, or a phone call.
Any similar inexpensive strategy can be all that's required.
An expensive marketing campaign is not necessary.
Look after any genuine prospect for your work just as well as you can.
You aren't doing them a favour by showing your works.
They do you a favour by looking and buying is an absolute complement!

Make obtaining follow up details a habit.
You do business with people and not find them again other than they choose.
Having contact details is the key to effective and low cost marketing.

Once someone is in your gallery or studio, get their contact details.
Find out what they like (your works, which, someone else's, whatever).
Even people who do not like your work (some) might know others who would.
That's why it's important to get a client's name, address and e-mail address.
Client marketing skills linked to a contact list are valuable assets.

This is why a gallery is reluctant to supply client name and address.
Even of people who have bought your works.
They want people they attracted at great expense initially to come back.
Not unreasonably, they do not want to see them move on to you.
You should accept this as being quite reasonable.

Don't just sell the work in question, discuss the future too.
Talk about when might be a good time to contact the client again.
Introduce other products or services that link with yours but you don't supply.
Other artists, picture framing, valuations of artworks, or whatever.

If the client wants something then say you'll obtain it for them.
Get as much as possible from a prospect and contact the supplier or artist.
Find out what they have that the prospect might want.
Make an arrangement with the supplier so you make money for your trouble.

Provide prospects with what they want and they're your clients.
Don't miss opportunities to obtain clients who keep coming back to you.
Then you'll have a dependable, perpetual, lifetime, income source.
Collectors can make enjoying an art career a realistic proposition!

Do you count the numbers who come into your gallery or studio?
A particular business made five sales from every fifty people who came in.
They'd run another ad, make another five sales, and so on.
But they never did anything with the forty-five people who didn't buy.
Let's say these were the figures for people coming into your studio.

The first thing you should do is analyze why those people didn't buy?
Some would be 'lookers', they love art but never actually buy any (artists).
Others wouldn't buy as your work is too expensive, complex, or intimidating.
On the other hand it could be for the opposite reasons to those.
Perhaps they just didn't like the sales person (you?).

Except for the 'lookers', it doesn't mean the rest didn't want to buy.
It just means you probably can't make that sale.

What if you pass these people to those with what they might want?
Then the prospect would be more likely to buy.
You could do this if you had an arrangement with other suppliers (artists).
You share the profit made on any sale to people referred by you to them.
What you earn from this arrangement is money you'd never otherwise see.
BUT now they are the other supplier's clients still better than nothing.

4. Asking questions can handle any sales situation.

Remember let the client answer each question before asking another.
BUT let's look at different kinds of questions and how they might help us sell.

Closed questions narrow the focus of the response.
They tend to encourage factual answers, often simply yes or no.
Usually closed questions begin with words like did, do, can, where and are.
Do you like my painting?

Open questions tend to encourage discussion.
That's because they do not have simple responses.
They encourage opinions as answers and often begin with how or what.
How do you think you'll explain your choice to your husband?

Permission questions open up a situation.
Used when you want to start deepening a client's perception of a problem.
Do you mind if I ask you about the paintings you already own?

Best / least questions are used to get a fuller picture.
What's the best (or worst) thing about owning artwork?

Magic wand questions explore what the answerer wants or desires.
If money was no object, what would stop you from buying that painting?

Catchall questions build bridges to the next phase in the process.
They start the transition to understanding.
Do you mean something like this painting?

Fact finding questions are specific and only need a short response.
Often start with what or how.
How will we deliver your new print?

Feeling finding questions explore feelings and motives.
How would you feel if I let you take the work home on approval?

Paraphrasing checks understandings and keeps things moving.
Do you mean take it home on approval?

Could your website ask questions?

5. How can you create better-performing marketing?

These days it's hard to get any message out there!
There are just too many alternatives competing with you for attention.
You need to increase the personalization of all your marketing.

An IMPORTANT marketing strategy is a feeling of talking directly.
The talk is to the reader or viewer and only to them.
Your information is relevant to their business or personal life.
This is the power of personalized marketing!
So people to do business with us, a "know, like and trust" factor is in play.
Depending on what you're selling the degree varies.
Prospects and clients need an insider's glimpse on why we do what we do.

How much of your life you want to broadcast is up to you.
But sharing at some level is important.
Share stories and things that happen to you (relevant to your market).
Help build the, know like and trust factor, so personalize your copywriting.
Today's technology makes generic, "Dear Reader" type marketing obsolete.
At a minimum, personalize copy by simply dropping the recipient's name in.
But there's more, including adding personalization to offers and guarantees.

Personalize so it seems as if a client is the only one receiving an offer.
Include specific copy based on their buying history.
Perhaps you know something about them (e.g. the type of art they own, etc.).
Which do you think would get more attention and a better response?
"Dear Art Lover, we help keep your art insured with our no risk art lovers' policy."
OR
"Dear Mike, we can help you keep your latest painting The Old Inn insured against accidental damage with our policy formulated exclusively for buyers of my work."

Obviously the second example is much more relevant.
That's because you are talking to the client about their specific purchase.

This example includes specific personalized elements.
This puts the burden of using this data on you, but if started, it's not difficult.

Personalize with photos.
Studies have shown photographs are one of the first things people look at.
It could be an advertisement, letter, or marketing piece!

Photos are an important way to connect with others.
Tell your story and create a 1:1 connection to support personalized copy.
Fun or interesting photos of your workplace, employees, family or vacations.
Again the goal is to create a connection with your market.

With the example (above) how much more powerful is it with a photo?
It could even be an image of the work purchased!
Whenever you use photos in your marketing, make sure to include a caption.
Captions are the next thing readers' look at after personalized photos.

Personalize with comics, cartoons, and caricatures.
Unique graphics are underutilized in most business owners' marketing.
There is an opportunity to connect for the same reasons photographs do.
Humour can be used wherever you want to create an emotional connection.
Comics, cartoons, caricatures in direct mail pieces, emails, and web sites.
And don't forget to use personalized elements in your captions!

Personalize with handwritten notes and thank you notes.
People complain of too much email, junk mail, and irrelevant marketing,
NOBODY complains of receiving too many handwritten or thank you notes.
Handwritten notes can be real handwriting or simulated handwriting fonts.
They offer artists a unique opportunity to connect with their readers

Steer people to personalized and customized web sites.
They're created specifically for an individual reader so personalize and profit!
A strategy should be congruent, meaningful, and real to help your career.
It should also help improve the lives and businesses of prospects and clients.
Personalized marketing strategies must be used in real and authentic ways.

In order to make a human connection personalized marketing offers.

It must not appear contrived.

The goal is not to use personalized marketing for the sake of it.

But to make marketing appear as real, natural, and 1:1 as humanly possible!

6. Testimonials can be very powerful and persuasive!

The best advertising is word of mouth isn't it?
Unhappy people are more likely to talk of an experience than happy ones.
Around nine times as much and that's word of mouth you can do without.
People who are happy mostly just don't think to tell their friends.
They simply enjoy their artworks, or whatever else they bought.
Ask them to make referrals and remind people they can help us.

Testimonials establish credibility with potential clients.
Your prospects will be more impressed by what people like them have to say.
Than anything else on your CV even winning a major award!
An effective testimonial can convince your paintings are what you say.
If you already have some positive feedback from clients.
Ask their permission to include the comment on your site or in your mailings.

Testimonials should also have a first name, last name, and location.
That helps to prove the recommendations are coming from real people.
I've only had one person who wasn't delighted to provide that permission.
That person was pleased but wanted the permission in an academic fashion.
I'm not writing for an academic audience so I didn't use her reference.

To encourage new testimonials, add a link to your site.
Attach a form that allows clients to give you their vote of confidence.
For example, "Click here to tell us what you think!"
Create an auto-responder that contacts your clients after they've purchased.
Ask them how they're enjoying their new work.

When clients send glowing praise in a letter or email contact them.
Ask for permission to add it to your site or mailing (includes email).
Do this even though you may not know exactly where to use the testimonial.
There's always somewhere that's appropriate.

If you've good testimonials, include them on your site or in promotions:
Put your best testimonials on your homepage AND in mid-sales copy.

It's most effective to put testimonials on each page of your website.
Particularly if they are relevant to what is on that page.
A page on your site featuring only testimonials is less effective.
People can skip this page or many of the testimonials.

It's that simple!

7. A client is anybody who wants to spend money on your works.

However, there's another way to look at this idea.
A client is the proper client for my artworks or artistic services.
What does 'proper' mean?
Whatever it means, not everyone willing to spend money meets this criteria.
Why have such, when getting anyone to spend is difficult.
Will this apply to all of your works?

Are some clients more important than others then?
I can remember the first painting I ever bought.
My wife and I attended an artist friend's exhibition.
He put this on himself, and we went to offer him support as 'artist friends'.

From this it follows that we had no intention of buying anything.
Well we did decide to buy one work that we particularly liked.
Not having a cheque-book, we paid a small deposit and went on our way.

Shortly after a capital city art gallery proprietor visited the exhibition.
She wanted to buy 'our' painting.
However, our artist friend had made a deal with us and couldn't sell the work.
From the artist's career viewpoint, who was the proper client for the work?
Was it us, just two friends, or the important gallery owner?

There is no doubt she was the proper client.
The reason for this exhibition was to attract the attention of the gallery owner.
It was his strategy to become a part of her team of artists.
There is no question the artist would rather to sell to the gallery owner.
It turned out well for our friend as he did become one of that gallery's artists.
But it could have been different.

I guess this story illustrates the professionalism of our artist friend.
He knew who the proper client was.
This knowledge had guided all of his thinking, except when selling to us.
Even then he couldn't know her response to the work would be similar.

How can we make sure the right people buy our works?
We need to test potential clients to ensure they are proper clients.
Sales people can do this by asking questions and so can you.
What sort of work do they own now?
What are they looking for?
What styles do they like?
Do their answers match your ideal client?

But also ask where do they live?
Do they entertain frequently?
Where are they likely to hang your work?

Ask questions, seek answers likely to lead to the profitable outcome.
A gallery owner can sell many of your works, so they can be profitable.
People who are leaders in social or professional groups can generate sales.
Wealthy people can buy more than those of us on more modest incomes.
That's not just because they have more to spend.
They also have more wall space and even several houses or offices.

Test your artworks and services so they are right for such a client.
For instance the price should be right - not too high, but not too low either.
How do you test this?
Well you could phone some past clients and ask them for their opinion.
Those who are most like your proper client will be best.

What if you have a work the National Gallery might want?
Would you sell to someone without making an effort to sell to the gallery?
How about a sale to your local council?
What about a sale to someone who hangs the work where many see it?
Are they important potential sales?
Than someone of modest means who admires privately.

Write down exactly what sort of clients you would like to have.
Now develop a range of questions to test if prospects match your ideal.

8. The best clients for your career are the clients you already have.

You are not a stranger to your clients.
Provided you have treated them well they're likely to stick with you.
Clients know and trust the people they have dealt with before.
And you know where they are and what they buy.
The best investment return is spend time and money on computer client lists.
Start by contacting just those special clients.

You can train your eye to notice who those best clients are.
You can train yourself to see other important things too.

Where do your best clients come from?
Is it Yellow pages, advertising, friends' recommendation?
You see similarities in best clients, things shared like suburbs or professions.
And by looking outside your business, you can spot others like them.
These people are, after your current clients, your next best prospects.
That's a bonus for identifying your best clients.

Clients and prospects make it possible to have a professional career.
It will not matter how well you paint though!
Your past clients are the best source of new clients but they need help.
A referral website makes it possible to get and stay in touch systematically.

A client might be worth a great deal of money.
BUT unless some comes your way it will make no difference to how you live.
So the question is how much is a client worth to you and what is meant.
Are we referring to their last purchase, or their next one?

How often do people buy?
People buy petrol almost daily and smokers buy almost daily too.
On the other hand most people don't buy a house anywhere near as often.
How often do people buy your artworks?

How much do you make on each sale?
Obviously a gallery's commission has to be deducted from the sale price.
The frame is also a deduction.

What else has to be considered?
The cost of attracting a customer is one factor (part of gallery commission).
It costs less for later transactions that the initial one.
It's not a matter of building interest but tapping the interest that's there.
Fame is a good illustration of this phenomenon.

What can you do with this kind of information?
You can work out how much we can afford to spend attracting new clients.
Then pay agents, sales people, galleries, advertizing, accordingly.

It might pay you to give a sales-person ALL profit on the FIRST SALE!
Provided it's to a first-time buyer.
Offer a first sale discount of your entire margin to the **NEW** client instead.
Each of these can be contemplated if the profit on future sales is sufficient.

Here's how you can calculate the long-term value of a client.
The $ cost per client to get (obtain, capture) the client = A.
The $ return per sale = B.
The number of purchases by a client in a year (average) = C.
The buying life of a client in years = D.
Obviously this is more than just a single sale.

The life-time value of a client = (BxCxD) – A
If C and D are enough, it's possible to make little or nothing on a first sale B.
Yet still make a great deal in the long run.
Think about special deals phone companies offer to change to their system.

An example might help you understand this aspect of business.
A typical client might cost $450 to obtain.
Your average return per sale is $900.

BUT people buy 1½ works in a year on average.
A buyer keeps buying for 5 years before 'their walls are full'.
These are made up figures just to illustrate the long-term value of a client.
In this case the long-term value is $6,300.
Does this give you a different perspective?

Most artists have a transactional view of sales, think in single units.
This does not provide you with the full picture (so to speak).
Even a guess is better than not doing this kind of thinking.
Estimate based on your experience with other similar clients.
You need to keep the necessary figures for a period of time.
Then these calculations become reasonably accurate guides.

How much is the long-term value of an average client of yours?
What will you do differently now?

A long term plan is needed to maximize a client's value to your career.
It's too important to leave to chance.
Use a similar approach for other calculations.
Calculate the $ per enquiry and $ per sale figures.
Then you determine the effectiveness of advertising and other promotions.

9. Are you status conscious?

Status is about social position and is an inherent part of a society.
A distinguishing feature of a society is variety and nature of social position.
This is neither good nor bad, just a fact.
It's a fact we can turn to our advantage, if we are aware and make use of it.

A basic human need is awareness of social position.
It governs much of our behaviour.
As artists we all have contributed to the nature of how others perceive us.
The isolated nature of artistic practice means it doesn't fit social framework.

Artists are often characterized as individualists.
BUT it may only be their art making that's done that way.
So artists are allowed some social latitude that (say) accountants are not.
We can be unconventional if we choose, as that's part of our social territory.

But the commercial side of art is definitely a part of culture or society.
If we want to sell our products, we must come out of isolated studios.
For success we need to understand how others think and behave if they buy.

Understanding this psychology is the key to successful sales.
Every time people buy, they seek to satisfy a basic human need.
Being aware of social position is one of those.
All have a perceived rank and it is normal to want to move to a higher level.
Acquiring objects services or experiences linked to a desired level, does this.
In one social group, a Mercedes car is up a level from a Chevrolet.
Whatever the object, the want being expressed is the same.
There's a need to belong to a particular social group (such as artists).

How can we turn this knowledge into sales?
Status symbols are not necessarily fixed and can change.
But they are always objects, services or experiences.

We sell objects (paintings), services (lessons), experiences (art tours).
Effectively link those, with a high status person or thing.
Then their status can be transferred to what we are selling.
You must be consistent, but it really is that easy.

Status rubs off onto whatever is around it.
Endorsements are a powerful way to add status.
Bill Gates announced that he'd just bought 10 works, or a whole exhibition!
Bill's endorsement would do wonders for the status of your works.
It would also allow higher prices and generate sales of other works!

'The President has a Joe Bloggs too, you know.'
I've seen photographs in American art magazines.
An artist presents one of their paintings (or a print) to the US President.
Such a photograph is a powerful builder of status for the artist concerned.
Not only does the artist become more important.
Other clients can bask in the reflected status given by the important person.

So it can be worth giving an artwork to a prominent person.
Provided you obtain photographs of them with it, and preferably yourself too.
You also need permission to use those photos.
Sticking them in your album of memories does nothing for your sales.

The photos and endorsement is more valuable than money foregone,
Use them in a newsletter, send them to 'International Artist'.
Have the local paper publish the photograph.
Put a print in every Christmas card you send, or whatever else.

Do all of these things if you can.
BUT as many people as possible see you, your work and the famous person.
The prominent person must be high status.
The famous person is sports person, TV star, politician, 'celebrity', or anyone.
But they need to be well known to add status to you and your art.

They could even be one of your relatives!
Provided they have high status they just have to be linked to your art.
All you need is the evidence and to promote the link.

They don't even have to be an art collector!
It's even possible they might not like your art!
That doesn't matter, provided it doesn't become public knowledge.
Are you going to tell anyone?

Listing people and collections having your work is a small step.
BUT it's unsophisticated with little influence except academics or curators.
A photograph of your work in the most prestigious collection is much better.

Even without the photographs use the information to convey status.
Listing owners does not do this, but selective quotes from the owners would!

You must promote the linkage for the status to adhere to your works.

10. Many products gain status through pricing.

Use price to move up the status ladder!
Designer jeans, wines, Rolex watches, Rolls Royce cars, and jewellery.
These are examples of status goods which are also expensive.
Generally original artwork is perceived to be a status product.
Some artist's works have more status than others and are more expensive.

So put your prices up, considerably.
Spend a great deal on superb frames justify the higher price as status.
If you've won a prestige art award it's easier than if nothing has happened.
Don't forget the famous people who were photographed with a work of yours.
They can add more status than winning an award!

So now you can see why artists should be conscious of status.
An increase in status can lead to an increase in sales at higher price levels.

Don't just put your prices up; link with higher status at the same time!
You can do this through endorsements.
I know one artist whose sister was one of Sydney's 'beautiful people'.
The artist quoted her comments endorsing her works.
People did not know they were related as they had different names (married).

One point often escapes potential clients for luxury goods.
They're excellent value for money.
Press releases emphasize how these goods hold their value forever.
Tiffany, Rolex, Mont Blanc estimate value of older, limited-edition products.

Their new ranges are instant heirlooms, for the evidence is there.
You can market your paintings this way too.
Clients need to be educated on buying up, not down.
The wealthy can do this, but everyone else should too.

1987 TAG Heuer introduced a watch with a band from the brand letters.
You could tell it was a TAG Heuer even when the face could not be seen.
This was the break through into Rolex territory for now they had a clear USP.
By 1990 TAG Heuer was fifth best-selling watch in the world by value.

That's why it's important to have a top frame on every work put on sale.
Like TAG Heuer band your work can be recognized by the frames used.
Some artists have specially made labels fixed to the frames, a similar idea.
The frames are **NOT** more important than the works.
BUT from a prestige sales view they're more important than artists think.

TAG Heuer is a prestige watch company in a very competitive field.
Rolex dominates but all market on a high price, limited distribution principle.
Each has to work hard to stay ahead.
Advertising in glossy magazines using well-known photographers is one way.

An alternative is to have celebrities wear one of the watches.
The Dalai Lama wears a Rolex.
People are highly aware of rank and social position.

TAG Heuer sponsors the Melbourne Grand Prix and other major events.
The watches are associated with the prestige that flows from the events.
They do **NOT** sponsor the local Under 10 soccer team.

All prestige firms have a range of promotional goods to give away.
They give them to prominent people who buy their products.
You could easily create small drawings used as gifts for high status clients.
Then you'd be just like TAG and other prestige companies.

Tiffany sells a wide range of prestige products.
BUT all are wrapped in a light blue box tied in a big white ribbon.
It's the same idea as people are paying for the package due to its status.

You can benefit from this phenomenon too.
How are your works packed for delivery?

People actually pay money to have presents gift-wrapped!
The appearance of a gift at the moment of presentation signifies status.

There is another way to enhance products in the prestige market.
Have them prominently shown in films and TV production.
James Bond has promoted Aston Martin and also BMW cars in this manner.
BMW paid a great deal for the exposure, but Aston Martin has fought back!

Not long ago Ford arranged a deal with a TV police series.
They agreed to supply Ford cars which were only used for the 'goodies'.
The bandits, crooks and others made do with other makes (perhaps BMW?).
This is could be beyond your resources but what about your imagination.

Place your works where they receive a great deal of exposure.
Selling to people who display your work publicly is worth a great deal to you.
People who only look at the work themselves pay full price!

A business foyer sale is better than a similar one for a private home.
But if the people entertain and move in the right social circles it is different!

Elite fashion houses are established in Sydney and Melbourne.
Tiffany, Gucci, Chanel, Louis Vuitton, Armani, and Versace are in proximity.
Concentration of prestige firms gives prestige to a common location.
Art galleries in Paddington (Sydney) or Toorak (Melbourne) illustrate this too.
If you want to be in the art status business that's where you have to be too.

You can't make prestige sales selling from your door in the suburbs.
The buyers just won't come.
You must go to where your intended clients are going to be.
That's why many galleries can do better than most artists!

Toyota entered the prestige car market so built a brand from scratch.
The word Toyota isn't on a Lexus (short for Luxury EXport to the US).

It has its own up-market dealerships.

There are marble showrooms where well-dressed executives help you buy.

Lexus has its service centres, 24-hour helpline and client service programs.

The Lexus project was researched before a single car was built ($3 billion).

The car was pitched at executives interested in innovation and prestige.

It was also cheaper than comparable Mercedes, BMW and Jaguar models.

BUT still by most people's standards expensive.

Lexus is now well established in its intended market.

If you are serious about getting into status, then treat it seriously.

The rewards are great but only if your homework has been done.

You'll only get one shot, so it has to be right.

It's all about positioning yourself and your artwork for your intended market.

Do you have a plan for moving further up-market?

Which means either, get one and implement it, or dream on.

11. Is word of mouth advertising best?

Many people talk easily fluently and frequently.
They are just the people to spread the word about your new exhibition.

A nightclub opened full and stayed full with little marketing.
They had a party for all the hairdressers in the city a few nights before.
There were queues on opening night.

Couldn't you do something almost the same?
Remember all they have to do is talk about your exhibition.
Their customers will do the buying.

Who are the talkers?
People who have clients like your clients and who also meet them regularly.
Theatre people could fall into this category.
They are usually talkers and often theatre-goers are also art buyers.

Target networks of talkers, particularly to reach hard to get people.
Service clubs, Rotary, Apex, Lions, BPW, or View Club are people networks.
They also meet on a very regular basis.

Non-English speaking people rely on advice from community leaders.
They could be doctors, solicitors, teachers, priests and the like.
Often they are asked for advice quite outside their areas of expertise.
If you get a list of people like this in your area, supply them with information.

Find out what talk networks of your present clients belong to.
Ask them which organizations they belong to.
Find clubs, industry groups, sporting bodies, religion, political affiliation, etc.
What talk shows do they listen to?

People who are not your clients belong to the same or similar groups.
Link up with them and offer to talk to these groups.
Many are always looking for guest speakers.

Do you remember you client's name?

Doing this can lead to favorable 'word of mouth' publicity, as they tell friends. How many different ways do you use to generate 'word of mouth' publicity? Which means you can measure the result to see which is working the best!

Chapter Two: Website that works.

1. How does YOUR website work NOW?
2. You might have to be a revolutionary!
3. Where does your website fit into this scenario?
4. Any website must be easy to use.
5. The stuff you write on your website is copy.
6. Off-line generators help your website pull its weight?
7. Create a special website.

1. How does YOUR website work NOW?

Artist's websites usually say to a visitor, here's my studio, have a look.
There are images, photographs and words, which include archival material.
But that's really a waste of a valuable opportunity.
All it does is make you (or anyone else who does the same thing) feel good.
Surely you don't want a website, just so you can say you have one?
In that case you won't worry if people visit and you don't know who they are?
Because you'll never want to contact them again?

BUT shouldn't your web-site be one of your marketing tools?
Then you will want people to visit and you know who they are?
That's because you'll definitely want to contact them again?
Like any aspect of marketing it should have some definite intended purpose.
If it doesn't, how will you ever know if it's any good (other than aesthetically)?

So just what do you want a website to do?
Write down why you need a website for there could be several reasons.
Does that mean you could need several websites?
Aimed at buyers and those interested in your paintings and prints.
Aimed at car collectors for your paintings and prints of this subject area.
Aimed at art galleries and agents who might help you sell your work.
Aimed at artists and potential artists because you teach.
Aimed at potential art hirers.

Having different websites makes sense.
Several websites separate different styles of work, medium or subject matter.
But the most important determining question is really, who are you aiming at?

Well who are you aiming at?
For new clients, what you write and photograph is different than for galleries.
So you could even have five different websites.
There will be overlapping of course (same work or photograph sometimes).
But generally each website should be focused on who it is intended for.
Do not worry at all about anyone else.
Market by using a rifle to hit the target, rather than a blunderbuss!
Decisions linked to a website make a difference to what you write and show.

Why have separate websites, wouldn't different pages do the job?
Do that if you don't mind potential clients knowing what you say to galleries;
Even then it's a distraction from what you really want to say to that person.

So what do you need anyway?
The first screen heading or major photograph is the key to a website working.
The first sentence or image is the ad for the rest of the website.
If this doesn't grab the people you want, then the website will fail to do its job.

Write lots of short sentences to interest intended website visitor.
See which comes up best which still might be the one you have.
BUT if the second sentence is 'But I sell like crazy!' and this is true.
You'll definitely have gallery owners, looking at the rest of the website.

Some artists think an artist's statement is needed?
If it is, then who would be interested in it?
The only people who worry about them are at universities or public galleries.
Maybe other artists are interested.
If this is your target, then go right ahead.

Otherwise my focus would be on 'What I can do for you!'
The "you" being the intended visitor to that website.

Same goes for a biography.
It depends on who you are targeting, as well as what's actually written.
Most artists are writing a sales document to attract possible buyers for works.
An artist's statement or biography is academic writing **NOT** sales stuff.

You could have a website aimed at your clients.
These are people who have already bought your work.

As purchase follow up biography and artist's statement are worthwhile.
But even then think about how it can become a re-selling device?
Also think about whether it should be on your website or paper!

What is your focus?
How does artwork help someone do or achieve something, or get pleasure?
This isn't easy, as most artists aren't used to thinking like this.
Keep the person you want to respond to the website, in mind all the time.
You are the last person to consider!
Can they get something as a result of buying one (or more) of your works.
If not they will not buy - it's that simple!

Look at ads on TV.
When you see an ad for Coke, imagine it's about one of your works.
Instead of a bottle, there's the painting.
Instead of 'Coke is it' people are saying something about one of your works.
Try the same thing with lots of ads.
Eventually you discover some that make sense so think more about them.
Now develop your website from this thinking.

If you are not sure what to do, just ask questions on your website.
Actually asking questions could be a very good way to go.
It's not threatening, which making a statement will be to someone.
Questions means a viewer contacts you with an answer, to move forward.
Their answer provides clues as to what they want.
The more they tell why they're interested in your work the less you need ask.

Some artists feel their artworks should stand alone.
They are very concerned about what to write.
A solution is remove most of what was written, leave works and photographs.
Add questions for people to respond to?
It's hard to go wrong with questions.

Here are ideas for questions some of which you might like to include.
What kind of background do you think an artist should have?
What is the post-code for where you live?
What background do you think someone painting images like these has?
What do you think I am trying to say with these works?
If an artist's work was about 'multiple dimensions', what would it be like?

How many artworks do you own?
Could you label the works on this website using simple words?
Abstract, figurative, portrait, landscape, imaginative, or word of your choice.
Art combined with mathematics, science, geography, music, or whatever?
How would a chemist's artwork be different from a geographer's?

Do you buy artworks?
Does an artist need to tell you about their work?
Do you get pleasure from the artworks you own?
What difference does knowing an artist's background make to liking their art?

Just write down lots of questions.
Turn all your statements into questions and re-write them again.
Keep going until you are happy with those you have for the website visitor.

Naturally contact details should be repeated, in any website you have.
BUT you should have differences so you can tell where someone came from.
Gallery + abstract site might be different e-mail address to client + abstract.
Same for others or you'll never know what sort of interest you are attracting.
Repeat contact information regularly through the website (even each page).
THEN as soon as a visitor thinks they'd like to contact you, they can.
Once at the end is not enough!

How do people find out about the website anyway?

There's no point having a great website if no-one knows about it.

However you do it you'll obviously need to promote each web-site differently.

You could advertise your art class website in the local newspaper.

But to reach galleries you may have to use a national magazine.

On the other hand car collectors could be reached by an e-mail campaign.

This could be through the worldwide network of specialist car clubs.

Does your website earn its keep?

2. You might have to be a revolutionary!

Artists who battle for sales may have to travel the revolutionary path.
They can't keep doing what they have been and expect a different result!
If you do that what happens to your past?
It's a new career, so what do works belonging to the old career do now?
Nothing except undermine your new status.
So you must distance yourself from those works.

The best way is to do it publicly and dramatically.
Find all your old paintings – the lot.
You no longer need them as insurance in case something goes wrong.
There is a momentum with your new work and the old ones do not belong.

As a professional artist your paintings are equivalent of actual cash.
To keep value all works must be your best so old ones can't be given away.
Look carefully at the old works, are any as good as the latest works?
If you think there are put them aside, but be objective and single-minded.
Your future depends on **NO** single work that's not your latest standard.

Plan a party where you live and invite your neighbours and the press.
Take the lesser works out into your front yard, or street if necessary.
Choose a fine week-end morning when there will be plenty of people about.

BURN them along with warped and damaged stretchers and frames.
If people ask do you have a fire permit – reply by saying what you are doing.
If the police come because of the fire – tell them what you are doing too.
Invite them to view your website as well.
Hold up each painting dramatically before throwing it on the fire.
Pretend you are a priest of an obscure faith and each sacrifice has a ritual.
If a spectator wants one free – refuse.

You're celebrating a turning point in your career as an artist.
You are burning your past.
Tell them about your coming website and give them the website address.

Set up a table so you can sign images of one of your new works.
This work is prominently featured on your new website.
The fire burns out, sweep up the ashes and put them in a special container.
Gather your table and the images and move to your house.
Without a glance backward wave to those still gathered at the scene.

You are on your way to fame and people will look at your new website.

I remember years ago reading in national papers something like this.
A well-known Australian artist did something similar.
He certainly burnt his old works.
I've never forgotten this for it's a dramatic statement in a field like ours.
Every work is assumed to have magic properties due to the creative process.
Use this common myth to advantage as you embark on your new website.

3. Where does your website fit into this scenario?

You want a website that works.
If it's not going to earn its keep, then why have one?
An answer might be interest visitors in your work if they haven't seen them.
You wish to remind them of a work they have seen in the studio or a show.
Maybe you want them to refer the site to friends, keep your name in mind.

It's best for prospects to visit a studio or gallery to see "live" work.
But many people live quite a distance away and such a visit is unlikely.
So is a website an answer to this problem?
The only reason **NOT** to invite prospects to visit is if it would turn them off."

So why would YOU have one?
Most artists' websites run the risk of trying to do too much.
But end up doing nothing at all for the focus is usually the artist concerned.
They try to cover everything dealing with that artist.

But most artist websites are ineffective.
Because just about everyone who looks at a site has already heard of you.
People use the website instead of coming to the studio.
They even use the website to avoid coming to your studio or gallery!

That's because they are basically set up as an electronic studio/gallery.
Why would people want to make a studio or gallery visit other than to buy?
Decisions are made on the basis of the website presentation.
So the website IS a sales tool whether the artist knows this or not!

But what about those people who surf the net?
The people who surf the net looking for artists want to sell something!
Or they want to involve you in some scheme or other - do you want them?

Could a website be a marketing tool of some kind?
Not perhaps as a way of selling paintings.
But for people acquainted, interested, and encouraged to follow-up in person.

Which means, you need to target those who know you, or know of you?
In your case have images of whatever you specialise in (your brand) etc.
For a landscape artist why not show the real thing, like a tourist brochure.
You could also suggest this is what you love and why you paint in this area.
Want to see some paintings – visit my studio.

4. Any website must be easy to use.

That means it has to be simple!
A website is just like any other part of your art business.
It must pull its weight, or earn its keep in economic terms.

A website is a relatively new marketing medium.
It is a challenge to get maximum value for time, effort and money expended.
There are opportunities, for seemingly impossible levels of success.
There are challenges which can mean expensive waste in other instances.
It's usually felt a website should make visitors want to stay and spend time.
Give information, make a site attractive and easy for visitors to find their way.

BUT if this is all that is done then you are encouraging browsers.
Earning its keep is more important than most other factors.
Your website needs to create credibility and rapport with the visitor.
They feel comfortable if you don't try to do everything and confuse people.

There are many reasons why people leave a website after an initial visit.
Not everyone buys, or obtains all the information they need, on their first visit.
They want to check the opposition or wait until they can afford what you sell.
Or even until they're ready as they don't need what you have right now.
Others may not fully trust you just yet, and want to talk to other people.

There are so many websites too.
It's difficult to remember how to find most of them.
So a visitor will probably need help to remember you.
That's why it's important you make it easy for them to return in the future.

The first step is to obtain their email address.
Then you can remind them who you are, what you offer and how you help.
Perhaps offer a free newsletter at your website (about a page).
Then at regular intervals remind people to come and check the latest issue.
Variations can be 'tip of the day, week, or month', or a monthly joke.
They return to check your website at regular intervals for new information.

Just making an offer on a website does not guarantee return visitors.
It doesn't guarantee return visitors without a way of reminding them to.
Thus a newsletter (not on the website) could remind people to come back.
They get the 'tip of the whenever', an important article, or a latest promotion.

Many websites are sales devices, whilst others provide information.
If yours is for selling then that's what it should do.
Site experiences enhance desire to buy without being pushy or over the top.
It should not be hard for a prospect to order, or take what the next step is.
This is not actually easy to do.
There are small things that need to be right before site traffic turns into sales.
It goes without saying that your client service should be spot on as well.
If buying on the net they know of non-arrival of goods and are nervous.

So make offers on a regular basis.
Never make a complete sales presentation in an email – just tentative.
Provide URL link to a suitable place in your website.

Back-end sales are those that follow the first sale.
Back-end sales are essential and profitable - Want the French Fries?
So you should, in marketing terminology, work the backend.
Other goods and services are available to clients (now) after the first sale.
Treat a client properly with excellent value and service on an initial sale.
Then the backend is where you can usually make the most money.
At least some trust has been established, which is a basis for future contacts.

Backend sales are profitable, as cost of gaining the customer was met.
There's no advertising, or other cost normally involved in finding a customer.
It should be the most profitable part of your business, no matter what is sold!
Payment for lessons (written) is a profitable back-end.
Buy a print, drawing, invite to an exhibition, are back-end offers if after a sale.

Your first sale is not just the costliest but it's also the hardest to make.
The chance is it is least profitable too, and not just because of the cost factor.
A first sale is likely to be tentative, "dip the toe into the water" exercise.

Someone's first purchase of your works will be small and inexpensive.
That's why you need such works to make that first sale!
But there's no money there is there?

But you need make little upfront if the back-end is sufficiently valuable.
Subsequent sales are likely to be higher priced and also more frequent!
Link to a lower cost per sale and you can see why you "work the back-end".
That's where the money is!
Collectors can keep you in business for years.
Now you have credibility – so there is less need to emphasize this.
Because you have credibility repeat sales are possible.

Getting traffic to the site isn't easy, either.
Search engines can generate traffic from those who don't know you.

But getting qualified visitors is more difficult and more critical.
These are people who might need whatever you have for sale.
It seems this is a process that never ends, but that's any kind of business!
Generally you'll need off-line promotions to take people to your website.

Can your website be a part of your sales strategy?

5. The stuff you write on your website is copy.

Copy is journalists' language.
Your website copy is advertising.
It's **NOT** a letter, a novel or short story, an academic essay, or invoice.
Basically you are enticing the reader to do whatever you want them to do.
You suggest a perfectly reasonable basis for them taking this action.

Now that's where many people go wrong.
There's **NO** attempt to sell anything.
Artworks are images described and presented in an academic fashion.
This is supported by a biography and a statement which is always academic.

BUT academic writing was never intended to sell anything!
Surprise – it doesn't!

BUT you want a website that pays so what should you write?
Put teasers in your copy.
Teasers don't tell the whole story but leave something to the imagination.
I'm going to tell you about (whatever) **BUT** first (something else)

Tell stories (from your life)
Potential clients just want to know what makes you tick.
Recount experiences … I remember when (whatever).
That's the time I almost … (whatever).
I remember when I was only young ….
Just fill in the blanks.
This makes you a real person not someone who seems different from them.

But focus on making sales – it is sales copy remember.
You are not just recounting your life story (do that when you are famous).
Link your story with reasons to buy.
Whenever you look at this work it will probably remind you of (whatever).
I had an experience like that back in …etc.

The entire front page of your website is a sales letter.
People then know upfront what you are about.
Include a guarantee, testimonials, benefits, etc.
If they are interested they'll keep reading.
The important part is the heading.
Then first paragraph, which expands heading.

The heading is the ad for the rest of the website.
A right heading attracts favourable attention from those interested in an offer.
Write for them and do **NOT** apologize to those not likely to be interested.

The first paragraph elaborates on the heading.
It also introduces the offer you are making.
The rest of the sales copy is then introduced and in turn gradually expanded.

Regular and relevant testimonials provide credibility.
Anything going to a client should have a testimonial including your website.

How long should your copy be?
Copy should be as long as necessary to do what it is supposed to do.
If it's too short you may not be able to tell the full story.
If it's long you may lose people who give up before the end.
But the people who give up were probably not likely buyers anyway.

Your copy shouldn't become boring!
It shouldn't be boring to people who are interested in what you sell.
It won't matter at all if it's boring to everyone else.

Possibly one long page is better than a number of short ones.
Each new page is a new opportunity to click out.
BUT people can skip sections, so the letter may not be read as crafted.
Build logical break-points built into a longer letter rather than new pages.

You might experiment with one long page or a number of short ones.
Find out which works the best for you.

There could be an introductory survey.
This targets people before registering or before leaving the site.
Then compare responses between those who register and those who don't!
This might tell if you are attracting the right people.
Once people register that survey should no longer be available.
Your results will be skewed and less valuable as a result.

How might you attract people to register for your site?
Having people register for your site is a key justification for having a site.
Get contact details of people interested in what you do is worth money.
How are you going to contact those people unless you have contact details?
There is no easier way to get that information than if clients give it to you.
Just knowing the number of hits on your site is useless information.

What can you offer to interest your prospect sufficiently to register?
People are not going to provide details unless there is some benefit doing so.
Possibly invitations to exhibitions, special deals, or other promotions.
Providing humour on a regular basis can be a powerful attractor.

But you don't just want anybody do you?
Lovers of different art from yours might like a newsletter on art investment.
BUT you are not likely to receive anything in return are you?
You must qualify the potential clients so they are the kind of people you want.
To do this you have to know what kind of person you want.

Generally they'll be people somewhat like your present clients.
Sort people to weed out tyre-kickers (who waste time but spend no money).
If registered, people can be contacted more frequently as they are interested.
They can be charged more for the same reason.

Finish with a call to action (what to do next) and a PS

If people don't like what you have to say or do – don't worry.
They're not likely to be clients anyway.

6. Off-line generators help your website pull its weight?

You need visitors to your website from somewhere other than the site.
That means you need off-line generators for your site.
This is what search engines promise.

One approach is lead generation off-line for sales on-line (via the website).
Off-line lead generators include; publicity and articles written about or by you.
They could be in the target market (magazines, papers, etc.).
Get your website address included in the coverage (more info see URL).
Send postcards – everyone reads them they reach people other mail doesn't.

Mail or email your contact list about your website.
Contact people with compatible products or services.
Articles published in magazines from various artist associations (for artists).
But decorator and collector magazines are better for buyers.
This approach is the kind of thing that is feasible for selling prints on line.
For selling original works a different method is likely to work better.

Another approach is also lead generation off-line.
THEN make the first sales step on-line (via the website).
AND conclude the sale off-line (in your studio or gallery).
Use similar lead generation activity as listed above.
The difference is how you set up your website.
This time the sale is visit the gallery or studio, rather than a particular work.

Customizing to fit
Include people's names and relevant personal information in any emails sent.
"So (name) if you (whatever), then (that will happen – the consequences)."
Automatic responses like this look more personal than "Dear art collector,"
Don't give much free hoping people buy it encourages buying free stuff.
Be upfront – we are selling (whatever) with a small amount of free stuff.

Don't give people too much choice:
Get their problem and then offer a solution.

Not choice of solutions for too much choice encourages inaction.
Whole computer systems are based on a very simple choice.

Sell one thing at a time.
Build your site by starting with a small one – sales letter + order form.
Build and generate traffic – get this working.
Later add components and develop the site more.
Sophistication can come by adding new sites and automation.

Focus on a specific aspect of your career and a related client group.

7. Create a special website.

Just imagine if your website had one main purpose.
To help people provide referrals to you and for your work?

Set up a website for gaining referrals from your best clients.
This could be a new dedicated website or your present one re-created.
The new one would be just for the referral role.
Either way your website is so clients can use it for this purpose (referrals).

As a second website this has a lot going for it.
It is better to keep the old site unchanged and build a new referral site too.
You won't want to make changes to a website you spent time and money on.
This website will only be a relatively small one because of its narrower focus.
It should therefore be cheaper to create than your original one.
Even if you use the same web designing process.

On a referral site there's no need for works other than your collectors.
Others distract attention and reduce their power as references as a result.
Photographs of actual places might increase the power of the referral.
You don't need prices as it's too early in a relationship with new prospects.

Have nothing except WIIFM (what's in it for me) for a potential client.
Maybe a special price for such people, or another reward for contacting you.
That happens as a result of the referrer showing them your website.

The website has works owned by recent and largest paying clients.
The works are at the owner's residence, office or where they are displayed.
The owner is next to the work in the photo with a comment by the owner too.
Permission to use that information can be gained at the same time.
There is no need for much else on the website except your contact details.

At the time you create the website.
Mention your desire to have a presence elsewhere to your featured clients.

This is repeated to the owners when the site is launched.
They could mention the site to any friends, associates or relatives.
Particularly those who might be interested in his or her recent acquisition.

Newly referred prospect live anywhere and download contact details.
But you still need theirs before they can be contacted.
If they are interested in your work they will get in touch.
Your website is set up to collect the new prospect's contact details.
You need the name of the referrer too.

Only provide your website address to people who have actually bought.
Say something like: "I don't normally give my website address out.
But as you like my work so much I'll make an exception in your case.
If you find someone who is interested in the painting you bought recently.
The one I photographed you with, show my website and they see it there too.
Provide the address of the website so they can look it up any time they want"

With the referral-based website all visitors will just look at the site.
There is nothing else to do.
The major seller is the referee when he/she tells someone about your site.
The site helps him/her do that effectively without seeming to sell.

There may be suggestions how interested visitors might contact you.
There is a way to capture names and email addresses, so you can follow up.
You can follow up by mail, email, phone or personal contact.
You know you are dealing with interested people who are likely buyers.
You also share a relationship with someone they know well, the owner.

To make actual sales:
Send images of selected work to a prospect by email or mail, or show works.
This is much more effective than just putting a number of works on a website.
Then hope someone falls in love with one and purchases as a result.
This rarely happens.

This is a website used as a rapier (hit a pinpoint target).
Rather than a blunderbuss (let's shoot and see if we hit something).
In our line of business this makes great sense.

Often there is a difficulty in overcoming skepticism.
Adding testimonials is one of the easiest ways to improve a web site.
A good one has more selling power than the best sales-copy on other sites!

Testimonials build trust:
Whether your clients rave about your works or the great service you gave.
They tell visitors that they had a positive experience with you and your works.
Testimonials aren't "salesy" for testimonials aren't written in your "voice,"
They stand out as candid and unbiased accounts of what someone thinks.

A good testimonial has the power to convince "tough sell" prospects.
Your artworks really made a difference in one person's life.

A client who shows someone the referral site is the BEST testimonial!
Let's say that you paint works which have a western (cowboy) focus.
A visitor is told about your website by someone else with that same focus.
If there are other testimonials they are written by people just like themselves.

Those are two good selling points.
The person who tells about your site.
And the people who have testimonials featured there.
They are likely to influence a buying decision in your favour!

That's the power of an effective testimonial:
It can convince your reader that your artwork IS special.
There is a magic in it that you can be trusted to deliver to your clients.

Don't forget your business card too?
That is a typical work and your signature on one side.
Your web address on the other.

Chapter Three: A referral website.

1. Getting things right!
2. A focused website is a more powerful website!
3. Would a loyalty website be a website that works?
4. What do you actually need?
5. Your collector website.
6. Expanding your market with a referral website.
7. BUT what could the artist also do without leaving New York.
8. So how will your next website be different?

1. Getting things right!

Do you need a shopping cart?

"What web hosting package should I have, one with a shopping cart or not?"
That's **NOT** where you start, but it's a question you need to answer.

Avoid serious hassles down the road.

So **DON'T** use a shopping cart that's provided by your web host.
Make sure you use a third-party or even a custom-built shopping cart service.
If web host and shopping cart provider are the same you totally rely on them.
What if your relationship with that company sours?
You set up your entire business infrastructure and ordering system again!

If you use a third-party shopping cart or install a custom-built cart.

The move from one web host to another will be **MUCH** easier.
Your business won't be tied into the original web host's technology.
Web hosts business is selling you as many of their services as possible.
Many of these services will add value to your business.
BUT you need to carefully pick and choose only the ones that make sense.

So should you create a special referral website?

At this stage of your career you quite likely already have a website.
Do you modify a website you have or create a new one for a referral role?
You'll be reluctant to change a website you've spent time and money on.

So a second website has a lot going for it.

This website need only be a small one because of its narrower focus.

It should therefore be cheaper to create than your original website.

Even if you use the same web designing process.

These days, you don't need to know a thing about HTML or web design.

You can get a great looking site up and running in no time at all.

The secret is to start building your pages with a template web site creator.

With a small amount of computer knowledge you can do it yourself.

There are plenty of website templates available.

Enter what you want without knowing sophisticated computing language.

The template even converts Word documents for you.

This is actually what many web designers do and charge you much more for.

There are an increasing number of website templates available too.

It is possible to buy a template for a simple website (more than you'll need).

And register a domain name for less than $100.

Does that give you a different slant on having a second website or a third?

Most basic packages include:

Domain name, web hosting, easy-to-use site template, and photo uploading.

There will be a number of domain-specific e-mail accounts too.

You can be up and running with your new web site in just a day or two.

Maybe depending on the length of their approval or acceptance process.

Template sites can't be customized as completely as regular web sites.

The margin for error is less than with an independent, do-it-yourself design.

You can get all your beginner errors out of the way at minimum expense!

Pay a little more a month to integrate a shopping cart into your template site.

You can opt a payment system of your own choosing to your pages.

PayPal is always free to sign up.

It's the easiest system for artists who are just getting started online.

Template sites are a great starting point.

Your site up and running and start experimenting with your online business.

As your career grows, you may need to develop a site of your own.
Then you can have more control over the look and functionality of your site.

It's never been easier to get started on the Web than it is today.

2. A focused website is a more powerful website!

Artists' websites usually run the risk of trying to do too much.
BUT in the end doing nothing much at all!

The focus is usually the artist concerned.
BUT the website tries to cover everything dealing with that artist.
That's why most artist websites are ineffective.
They are basically set up as an electronic portfolio.
Many cover the past as well as the present too.

Why do you want people to visit your website?
To get them interested in your work if they haven't seen it?
To remind them of a painting they have seen in the studio or at a show?
To have them refer the site to friends?
To keep your name is their minds?
The only reason **NOT** to invite them to the site is if it would turn them off.

Websites reach those who live long distances from a studio or gallery.
It is the world-wide web after all!
BUT many people live so far away a studio visit is unlikely.
Although it's probably much better to get them to the studio to see live work.

Just about everyone who looks at your site has already heard of you.
People use the website **INSTEAD** of coming to the studio.
Decisions are made on the basis of the website presentation.
Why do people make a studio or gallery visit other than to actually buy?

A website IS a sales tool whether the artist knows this or not!
This requires the artist to think about it differently.
Get people acquainted and interested, and encourage follow-up in person.
A website exhibition invitation could be included for those living elsewhere.
BUT omitted from the ones you could reasonably expect to come in person.

You need your site to target those who know you, or know of you?
You could have images of whatever you specialise in (your brand) etc.
A landscape artist could show the real thing, like a tourist brochure.
Also suggest this is what you love and that's why you paint in this area.
Want to see some paintings – visit my studio (or check my other website)?

Just imagine if your website had ONE main purpose.
To help people provide referrals to you and your work?

Then re-create your website just so clients can use it for this purpose.
Just have WIIFM (what's in it for me) stuff for potential clients (as referred).
Maybe you have a special price for them, or some reward for contacting you.
But only as a result of the referrer showing them your website.

You only provide your website address to people who actually buy.
When you do say something like:
"I don't normally give my website address out, but as you like my work so much I'll make an exception in your case. When you find someone who likes the painting you've just bought, show them my website."
Your website will be set up to collect the new prospect's address details.
As well as the name of the referrer.

This has to be better than other approaches designed to get referrals.
Like handing out business cards, or sticking cards on the back of works.
Those strategies do work, but they wouldn't hold a candle to this one!
There's a strong possibility such a website could pay for itself in extra sales.

Decide exactly what you want the website to do – be specific.
Make sure that is what the website does **AND** nothing else!
Selling is a part of marketing.

3. Would a website that generates loyalty be a website that works?

Is there a system for developing client loyalty?
Loyal clients, who keep buying your work, support you through your career.
Your career can be built on such people so it's important to remember them.
A system is a great memory aid.
Then you never forget these very important people.
Add value by helping your clients remember you in a positive way.

Know your client and find more like them.
Turn clients into your biggest advocates, they'll find others like themselves.
'Frequency of Interaction' principle and you can't contact someone too much.
Timing is the key to everything.
You do not know when the timing will be right for any particular client.
You just have to be persistent for sooner or later it will be spot on!

Find ideas to stay in touch regularly, and then do them.
Schedule a working frequency of contacts for your clients.
Test the waters before plunging in.
Is what you want the same as what your client wants?
Many artists maintain contact with clients by sending a regular newsletter.
This is a single page or more elaborate, but the regular contact matters most!

You might remember the 80/20 rule (Pareto Principle).
That rule should be applied to your newsletter.
Then you can be seen as a service provider rather than a sales-person.
80% of the information is about stuff of interest to your clients.

Then you can write 20% on yourself and your services etc.
Is your newsletter like this?

Conduct a seminar on anything you think could interest clients.
This is different from one based on what you know unless it interests them.

To find out what is of interest, just ask.
Eventually you might conduct seminars for clients on running seminars!
A similar idea is to have people as speakers (trade, suppliers, authorities).
But again the focus must be of interest to the audience.

Do you have loyal clients?
What can you do to make sure they stay that way?
Here's something to think about!
The son of an artist I know visited a client in another major city.
Her son mentioned the artist's website was on this guy's "Favourites" list.
He shows it to people who have any connection to where she lives.
Now just think carefully about that.

This same artist has had her accountant send several prospects to her.
They have been shown the same site in the accountant's office.
There are four of the artist's paintings there too.
You might have had similar experience.
Perhaps it may have happened and you were not aware of it?

Think of the artist's son's observation and the accountant's behaviour.
Do they provide clues about a whole new approach to websites for artists?

Just imagine if your website had one main purpose?
It was to help people provide referrals to you and your work?

Then re-create your website just so clients can use it for this purpose.
Have nothing else there, except what's in it for me stuff for a potential client.
That's the one being referred.
Do you have a special price for such people or a reward for contacting you?
This happens as a result of the referrer showing them your website.

You only provide your website address to people who actually buy.
Say something like: I don't normally give my website address out.
But as you like my work so much I'll make an exception in your case.
If you find someone who likes the painting you bought, show my website.

Your website is set up to collect the new prospect's address details.
You'll want the name of the referrer too.
The website should only present your specific focus - nothing else!

This has to be much better than usual approaches for getting referrals!
Strategies such as business cards, sticking cards on the back of works, work.
But they wouldn't hold a candle to this one!
There's a strong possibility such a website could pay for itself in extra sales.

Decide exactly what you want the website to do – be specific.
Make sure that is what the website does **AND** nothing else.
Selling is a part of marketing **AND** so is a website!

4. What do you actually need?

This site is for collectors to introduce others to you and your work.
But keep in mind it's an introduction.
Thus you only need the bare minimum on your website for it to do its job!

You will need:
A collector page, as that guarantees the referrals.
Your signature, a typical work (several times if necessary), contact details.

There's a trick here though, which contact details?
It will be wherever you want the new potential client to make contact.
This might be the address of your studio, or gallery, or email address?
But not all of these, just one!

Possibly some images of what inspires your work.
Photographs of a rocky coastline you paint often, or dogs, or whatever else.
But the photos are of the real thing, not your paintings!
The prospect has to go to the gallery (or your studio) to see those.
If your work is more conceptual, write a few words to convey the same idea.
But do not turn this into an artist's statement!
Be casual, write in the first person (e.g. I often wonder about … whatever).

Ideally there is a picture of someone famous with a work.
Just lay enough bait so a prospect just has to visit a gallery or studio.
There they can find out more and see the real works.

But definitely there is a picture your client (the referrer) with your work.
The picture shows there the work is now hanging.
They can find out more by visiting the client to see the actual work.

Your collector client can fill in some of the gaps.
That is way better than you having everything on the website.
Winnie the Pooh only had to smell the nectar to find the honey pot.
Art collectors are exactly the same!

So you DON'T need:

Anything at all that doesn't do the job you want this website to do.

If in doubt then leave out, should be your motto.

Leave out any images other than a typical, signature image.

Avoid different galleries showing variations of your work.

No CV is necessary and an artist statement is not needed either.

Don't write about the art at all.

There should be no photograph of yourself.

If they want to know what you look like the prospect must meet you.

You'll need to offer your referrers off-line tools to generate the referrals.

Provide postcards – everyone reads them.

A typical work and your signature is on one side and web address the other.

5. Your collector website.

One artist redeveloped her website in order to make it more effective.
Many things were removed, but one thing she included was a collector page.
Previously I'd used the idea as an alternative to an artist's CV.

She had shots of works hanging on walls of people who bought them.
In addition there were testimonials from the owners.
The testimonials could be written or spoken (maybe both?).

So instead of what is presently on your site.
Visitors might simply see a page or section of your site.
This has photos of installed paintings in the homes of some of collectors.
Their quotes would be there too.

Most would love to do this for you.
Will they show your site to other people do you think?

A video-clip might be used, provided it fits your website focus.
It's a collector website to breed loyalty from collectors of your artworks.
It would also help them generate referrals for you at the same time.
That could be all you need!

A suitable video clip could be used as suggested.
It might be better to follow up those who ask for it from visiting the site.
You could have one specially made.
In this situation it would pay for itself.

Your biography is for following up a sale.
It never makes a sale.
It is interesting to know about the artist but **AFTER** a purchase was made.

Most art buyers do not even read your biography.
At best they scan it.

Well who does read a biography?

One who always reads a biography is the particular artist concerned.

Quite naturally they are proud of what they've achieved.

Other artists look to pick gaps and inconsistencies.

Academics are interested for it's their currency.

So leave your biography out of the website!

Same goes for your artist's statement for your works are your statement.

They surely need nothing else.

6. Expanding your market with a referral website.

An artist living in New York wants to expand to Los Angeles.
Other than family he knows no-one in LA.
Anything can be done for it's just a matter of going about it the right way.

The artist concerned is successful.
He can afford to do what someone lower down the ladder might not manage.
But that's also why he wishes to continue to live and paint where he is now.
He is prepared to go to LA to meet people and attend functions occasionally.

Eventually he will need someone to represent him in LA.
He is looking for a gallery to take on this role.
That is not a wrong line of thinking but it is conventional.
BUT he may miss some alternatives such as developing his own agent.

Obviously our artist doesn't know the LA art scene.
He has no people contacts within its ranks so dealing with that is a priority.

On the other hand his family must know people in LA.
They may not necessarily know people with art connections (but they might).
Let's assume they do not have that kind of knowledge.
They could be a link between the people who need to be known and artist.

Some of the family resident in LA could do legwork for the New Yorker.
But what can they do?

It doesn't matter if the artist paints commissions or sells in galleries.
They could find contacts for him.
BUT how will they do this?

Just collecting names and addresses is a start.
They can buy and read the LA newspapers.

In particular look through the local news and social pages.
Make a list of those who seem to be the most important in Los Angeles.
Own businesses, CEO's of major companies, do good things, attend events.

Do it for a month and they have most key people from the newspaper.
Check the list against actual news pages, delete anyone who is in trouble.
The remaining people will get the chance to meet the artist and help him.

Now make lists of people who don't like publicity, but are important.
Relatives ask friends, family, people they know, people they've met, anyone.
They want **THEIR** list of influential people in the community.
They'll also tell them about these people – why do they buy art?

Sort the friends' lists into one list.
Ask their friends etc. why they have included each person.
If they can't say why – then cross them out.
Leave them in if other friends can say why.

Now combine the two lists (newspaper / friends).
Ask friends if they know anyone who knows one or two people.
On the combined list.

Ask the friend to call that person (who knows the important person).
Talk to them about anything at all in a short comfortable conversation.

During the conversation they should mention the NY artist's name.
They should also say that he is an extremely talented artist.
Mention the important person as someone who could be interested.

The artist's friends should note who will ring whom and relax.
The New York artist's word of mouth advertising has begun.

He will want his paintings to be seen after they've been commissioned.
When buyers are chosen he makes sure they're also good sales people.

The relatives are not finished yet!

They should write down everyone they know, just everyone!

Obvious possibilities are their friends, other relatives, and acquaintances.

They should not worry whether they will buy the artist's paintings.

After all everyone has to start as someone who hasn't bought (yet)!

Also they shouldn't worry about where they live.

Just write down whoever they can, even if the details are incomplete.

This exercise will mean an extensive list is already constructed.

Write down everybody they do business with.

They include all the people who depend on them for some of their income.

They should not worry at all about whether they're likely to buy art or not!

This'll be a lot of people too - more than they think when they first start.

Don't worry about whether all the details are there or correct.

They should just write down whatever they can.

The relatives have supported these people they can expect support back.

Butcher, baker, bottle shop owner, electrician, hairdresser, lawyer, doctor, dentist, car repair person, accountant, car dealer, insurance representative and professional people are included.

Depending on the number of family members this could be a large group.

They write down everybody our artist might hope to do business with.

Anyone they know who might exhibit the artist.

Anyone likely to have names and addresses of those interested in or buy art.

The artist will need the addresses, email addresses and phone numbers.

Picture framers, galleries; art shows organizers, publishers and others.

The relatives could research these using directories.

Possibly each relative could take one section.

The relatives do not need to personally know these people.

Obviously clients he'd like will need to be listed.
Anyone they know of who collects, or buys, artworks should be included.
Even include those who might possibly buy artworks.
Look in the Yellow Pages under headings 'Medical Practitioner', 'Lawyer' etc.
Check back to the White Pages to find home address (not all will be there).

Details of organizations who are major art purchasers are also needed.
Relatives remember major corporations, government, education institutions.

They might be able to write down other potential contacts for the future.
I am referring to spouses and gatekeepers.
Gatekeepers are secretaries, receptionists, assistants for important people.
They determine whether you get to see or talk to the person concerned.
Their co-operation opens doors that would otherwise be hard to penetrate.

These contact details and it's possible relatives won't need to do more.
They could now want to.
He has enough information to consider developing his own agent in LA.
That could easily be one of the relatives!

If they think the list is too big, they should think again.
There is no such thing as a list that's too big.
Just keep adding those names and details.
Again different relatives could do different aspects.

List quantity is irrelevant at this stage.
BUT it **IS** possible to make 10 x more with a small **BUT** well targeted list.
Not all prospects and clients are equal but here it is just a preliminary listing.

The focus is yet to come.
Eventually our artist will choose who he will allow to handle his works.
That is a key part of being in control of his career rather than drifting.
Names, address, emails, phone numbers are needed before culling begins.

They are also needed before any agent is approached.

Consider the reception he is likely to get if he can say:

I want an agent to represent me and my contact list has 5000 people in LA. I'll make them available to you if we decide to work together."

Compare that with:

"I'm a New York artist and I'd like someone to represent me in LA"

7. BUT what could the artist also do without leaving New York.

This artist is successful so he has plenty of clients.
Presumably their details are all on his database.
That contact information means there is quite a lot that can be done.
He creates a **NEW** website and uses it for a client base, including LA.

His first step is to contact his most recent and largest paying clients.
The letter, call or email mentions he is planning to develop a new website.
He will ask the clients if he could use the work they bought on this website.
Most will agree because they will be flattered to have their work featured.

They will be followed up for a photograph.
The photo has the work at the owner's residence, office or where displayed.
The owner is next to the painting and in the photo.
The artist chats to the owner and writes down any favourable comment.

He also asks for permission to use it on the site next to the photograph.
He assures the owner the site will be sent for approval before it is public.
This website will feature his work and the quote.

The photograph is of the owner in his environment with his painting.
It is **NOT** a photo of the painting!

You tell the owner he is planning to expand to LA. or wherever.
You casually ask clients if they know anyone in LA who could like your work.
The owner approves the photograph and comment.
Now add them to the new website.
There is no need for much else on the website except your contact details.

When the website is up and running:
The NY artist again contacts the owners and ask for their opinion on the site.
The works are at the owner's residence, office or where they are displayed.
An owner is in the photo next to the painting and their comment is there too.

Suggest they mention the site to any friends, associates or relatives.
Particularly those who could be interested in his work.
Again he might mention his desire to develop a LA presence.

This is a website for gaining referrals from your best clients.
Newly referred prospects can live anywhere and download contact details.
The artist needs theirs before he can contact them.

If they are interested in his work they will get in touch.
That's when he will ask for and receive those elusive details.
The artist will also need permission to use the contact information.
A prospect referred to you by an existing client is desirable.
This should be a focus of your marketing.

That's because earning prospect trust is a major barrier to sales.
But when someone makes a referral they obviously trust you.

A portion of that trust is transferred to the new prospect.
There is a degree of trust between his client and the prospect.
He has a better chance of business with the new prospect because of it.
They'll at least give him a hearing.

8. So how will your next website be different?

You may have an old style here's everything I do kind of website.
You'll be reluctant to abandon a lot of time, thought, effort, and money.
It is better to keep the old site unchanged and build a **NEW** referral site too.

The seller NOW is the referee who tells someone about your site.
The site helps him/her do that effectively without seeming to sell.
BUT there should be suggestions how interested visitors might contact you.

The referral site doesn't need works other than those of collectors.
Any other works distract attention from the message of the collectors.
Their power as references is reduced as a result.

Include photographs of actual places that inspire your artistic focus.
They increase the power of the referral experience.
BUT you don't need prices if you want to preserve client confidentiality!
And it's far too early in a relationship with new prospects.

With the referral-based website visitors will just look at the site.
BUT there should be a way to capture visitors' names and email addresses.
With their addresses follow up by mail, email, phone and personal contact.
So make suggestions about how interested visitors might follow up.

You know you they are interested people who are potential buyers.
You share with them a relationship with someone they know well, the owner.
Eventually you send images of selected works to a prospect by email or mail.
You might show works in person to arrange actual commissions or sales.

The most common goal for a first site is to create a virtual portfolio.
Collectors, dealers, and curators get familiar with you and your work.
It's like handing out a brochure they can look at anonymously and at leisure.
You hope then they might contact you and ask to see your work in person.
Most of these websites are far too long, without any real justification.
Stuff goes in just because you like it but is this good enough?

A website should help you achieve a particular objective.
It doesn't matter if it is long, or short, provided the content does its job.
Perhaps your priority was to develop an effective contact list?
But do more and more contacts delivering the results you want?
Would you prefer fewer but better contacts?

It is essential to have a list of prospects.
BUT it is most important to have the **RIGHT** list rather than a large one!
Consider your work and the kind of person it may appeal to.
Filter or qualify visitors so those most likely to be interested are attracted.

In that case what if there were NO images?
A website could have details of paintings (price, subject, size, medium).

BUT with **NO** image.
Would be cheaper (no image).
Could supply more (less space).
Could trigger prospect imagination (lead to commission).
Give details how to make enquiries (visit gallery, studio, phone you, etc.).

But what if you limit access to this page to your very best prospects?
Then your website could focus very strongly and with precision.
Your focus should provide likely opening lines.
You have the right people so you will not need to attract interest from them.
You will still need to take a great deal of care when crafting the opening line.
It will **NOT** usually be the first one you think of.

What is your focus for this website?
Focus isn't just about what you paint.

It's a guiding principle to being effective at anything.
Focus is why your new website is the opposite to a standard approach.

BUT your homepage needs an effective visual.
An image of the inspiration behind your works is an excellent visual.
That leaves an interpretation of a painting to your client's imagination.
This might be a racehorse, typical landscape, a face, whatever.

You need an image of a typical work.
Similar to many others that you have done and will relate to the inspiration.
It is owned by one of your clients.
Show it in the office, board-room, foyer, dining room, or wherever it hangs.
The proud owners will be with the painting.
It is the kind of painting you expect others will also commission or buy.

Selecting photographs:
Get some from any person or organization you'd like to have as a referrer.

Ideally the work should be in its normal site.
But you should select which photographs you will use for it is your website.

The photographs entice people to want to see and own your paintings.
It will not matter if some photos only have parts of your paintings showing.
There only needs to be just enough to whet the appetite.
But there must be sufficient for your referrer to be able to identify with.
Glimpses of portions of paintings (show your signature) should be enough.
Tantalize so the viewer wants to come and see the real thing.

Start collecting testimonials from your collectors.
When people say something you like.
Ask would they mind writing it down so you can use it in your website later.
Owner's comments add credibility to your website.
"I've been busy collecting testimonials for my website. Everyone I have
asked has been happy to help me out with a photograph and a few words."
(Richard Rogers Strathalbyn Australia)

In addition you need:
Your signature as it appears on your paintings.
Your name needs to be prominently featured.
Don't forget your contact details.
No artist's work sells itself and neither does a website.
People buy art just as they buy everything that is sold.
Visitors need guidance to find the right place where sales can take place.

Your site needs effective navigational tools as well.
The site will usually be small so the user will easily finds their way around.
Your client will steer the user, and help them find pathways around your site.
They will make sure they see what you want them to.

But what if you had a different website for each of your clients!
The basic site is the same but only **ONE SET** of collector images and quotes.
The cost is not great and time taken to prepare each site not great either.
A lot of the same stuff can be used on all sites.

Let's say each site cost $250.
Only one person (and you) has access to the domain name.
Say they pass that to 10 of their friends and associates.
Two of those buy your work and spend $1000 each.
You develop a website for each of them.
What if your works are $10,000 each?
Are you ahead?
Do you need a gallery?

Keep it simple.
The drawback to using fancy bells and whistles is they are distractions.
What is really important is the visitor to get the information they want, quickly.
It's a win/win.
You leverage the opportunity into something better for yourself and the client.

This website will revolutionize how people (not just artists) use a site.
Done right and used properly, it can be a rapier (hit pinpoint target).

NOT a blunderbuss (shoot we might hit something) of typical websites.

In our line of business this makes great sense.

It also makes search engines redundant.

WHERE NEXT:

**BUT being a professional artist is NOW harder than it ever was.
These books are on earning money from a professional art career.**

Make Exhibitions Work
http://www.amazon.com/dp/B0882MFPGX

Art Hiring
http://www.amazon.com/dp/B0884JWR2S

Agents
http://www.amazon.com/dp/B08847Y9KS

Courses and Workshops
http://www.amazon.com/dp/B0884B51JB

Selling Prints
http://www.amazon.com/dp/B08846SWQW

Retirement
http://www.amazon.com/dp/B0884D9TBP

Art School
http://www.amazon.com/dp/B08849FV59

Gallery Co-Operation
http://www.amazon.com/dp/B087637FFW

Selling Strategies
http://www.amazon.com/dp/B0882JH3WN

Copyright
http://www.amazon.com/dp/B0892HWYTV

**BUT being a professional artist is NOW harder than it ever was.
This book is the last of a series on earning money from an art career.**

TAKE THE PLUNGE and Consider a Gallery.
http://www.amazon.com/dp/B0874JF964
Hardback
http://www.amazon.com/dp/B09GQRB34T

N O T   N O W :

Perhaps one of these books could interest you then?

What about your own memories?
YOU could publish them – like I did!
http://www.amazon.com/dp/B087DWKPTP

A simple way to start developing creativity.
If you are a parent, teacher or someone who meets a group regularly?
http://www.amazon.com/dp/B088T1KFQZ

The way most people start to become an artist!
http://www.amazon.com/dp/B088Y1DPL6

About some more of my memories.
http://www.amazon.com/dp/B088Y4RPL9

SEND TO:

**Know anyone interested in chocolate recipes?
Send them a link then.**
http://www.amazon.com/dp/B0882HK9Q9

Know anyone interested in THIS book?
http://www.amazon.com/dp/B08846SWQP